wild, wild world

GORILLAS
AND OTHER PRIMATES

Written by
Clare Oliver
Illustrated by
Terry Riley

p

This is a Parragon Book
First published in 2001

Parragon
Queen Street House
4 Queen Street
Bath BA1 1HE, UK

Produced by

David West Children's Books
7 Princeton Court
55 Felsham Road
Putney
London SW15 1AZ

British Library Cataloguing-in-Publication Data

A catalogue record for this book is available from the British Library.

ISBN 0-75254-671-6

Printed in Italy

Designers
Jenny Skelly
Aarti Parmar
Illustrators
Terry Riley
(SGA)
Rob Shone
Cartoonist
Peter Wilks
(SGA)
Editor
James Pickering
Consultant
Steve Parker

CONTENTS

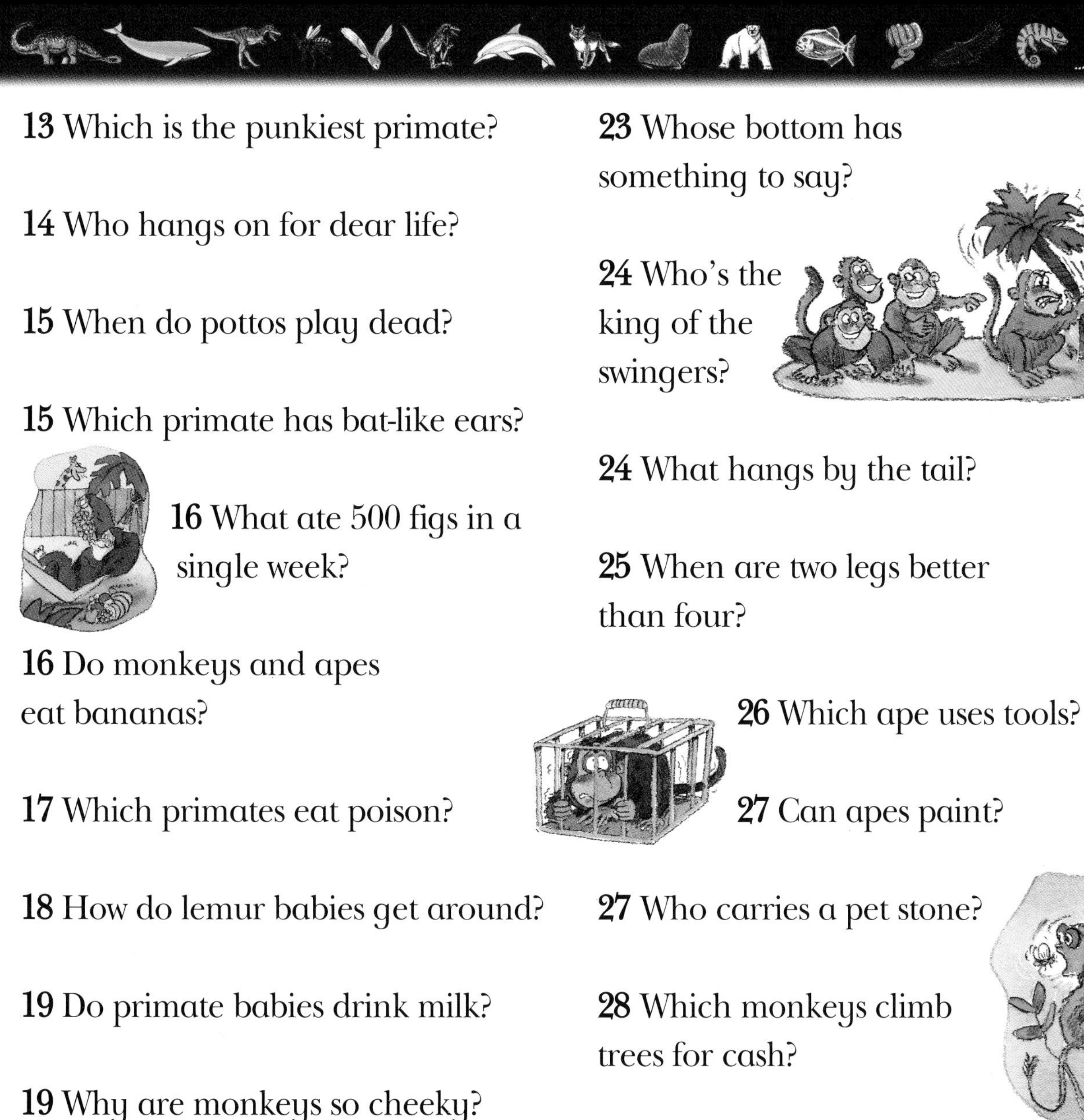

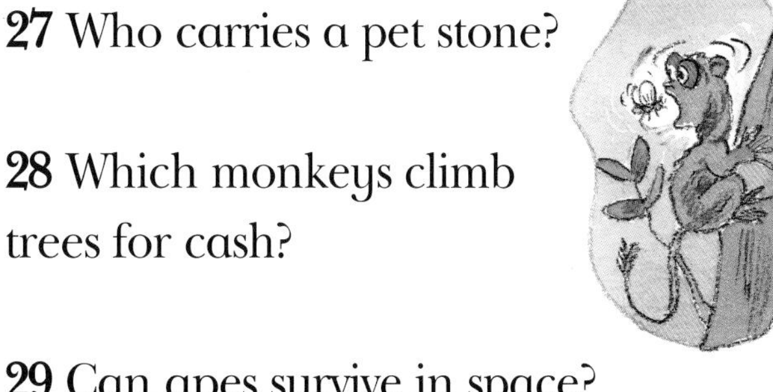

How do you tell a monkey from an ape?

By looking at its bottom! If it has no tail, it's probably either a 'great ape' – a gorilla, chimp, orang-utan, or bonobo – or it might be a type of gibbon or 'lesser ape', such as the siamang. Except for a few out of over 100 types, monkeys do have tails. Monkeys and apes belong to a group of intelligent animals called primates.

Amazing! You are a primate! Like all primates, you have forward-facing eyes, a big brain and hands that grip. Your genes (the instructions which tell your body what to be) are actually very similar to a chimp's genes.

Who's stopped being a primate?

Experts used to say that the tree shrew was a primate, but really they're more similar to insect-eating creatures such as moles, shrews and hedgehogs.

Slender loris

Are monkeys and apes the only primates?

Lemurs, bushbabies, lorises and tarsiers are all primitive primates. They have smaller brains than monkeys or apes and rely more on their sense of smell than sight.

Crowned lemur

Is it true?
Primates were around in dinosaur days.

Yes. People have found fossils of early, squirrel-sized primates that lived about 70 million years ago – about the same time that terrifying *Tyrannosaurus rex* roamed the land.

Which primates play in the snow?

Japanese macaques are sometimes called snow monkeys. They are found farther north than any other primate, except humans. For five months of the year, the mountains where they live are covered with snow. They grow an extra-thick winter coat and there are lots of hot springs, where they can take a steamy bath when they are feeling the chill.

Amazing! Brazil, in South America, is home to about 50 types of primate, more than any other country. Most of Brazil is covered in rainforest, which makes the perfect home for a tree-loving primate.

Which primates play on the rocks?

There are about 100 barbary macaques living on the Rock of Gibraltar, just off the southwest coast of Europe. They feed on grass, young leaves, spiders and treats from tourists.

Barbary macaque

Is it true?
Lemurs are found all over the world.

No. Apart from in zoos, lemurs are only found on the islands of Madagascar and the Comoros, off the southeast coast of Africa.

Chacma baboon

Which primates play in the sand?

In the Namib Desert, Africa, there's less than three centimetres of rainfall each year. The chacma baboons, which live there, eat wild figs for moisture, but often don't drink water for two or three months.

Which ape has the longest arms?

In relation to its overall size, the orang-utan has the biggest armspan. Its arms are three times as long as its body, which is just right for an animal that spends its life swinging from tree to tree! The name orang-utan is the Malay for 'man of the woods'.

Is it true?
Gorillas in the wild are bigger than gorillas in zoos.

No. Life in a zoo can make gorillas rather lazy and, sometimes, rather fat! The record-breaker was a male called N'gagi, who weighed in at a whopping 310 kg. That's about the same as five adult humans!

Amazing! The gorilla is the world's biggest ape. It is a tiny bit taller than a man, but usually about three times as heavy.

Which is the most colourful monkey?

Male mandrills, which belong to the baboon family, have very brightly coloured faces. Mandrills are also among the biggest monkeys, at just under a metre tall, with a weight of about 20 kilograms.

Mandrill

Mouse lemur

Which is the world's smallest primate?

The eastern brown mouse lemur of Madagascar is truly tiny. From the top of its head to its bottom, it measures just over six centimetres. It could easily sit on your palm, if it wasn't so shy!

Where do gorillas sleep?

It's not just birds that sleep in nests – huge gorillas do too! They bend branches in bushes and trees and make a cosy bed just above the ground. Sometimes they make a mini day nest, where they snatch a midday snooze.

Is it true?
Gorillas are monster meat-eaters.

No. Despite their enormous size, these gentle giants are vegetarians. They feed on fruits, roots and vegetables, especially delicious wild celery.

Which gorillas go grey?

Adult male gorillas are called silverbacks, because of the silvery grey fur on their back and face. The silverback is the leader, who defends the troop.

Silverback gorilla

Amazing! Gorillas use sign language! Wild gorillas communicate with grunts and body language. But a gorilla called Koko learned proper sign language, as used by people who can't speak or hear.

When is it rude to stare?

It's always rude to look straight at a gorilla. In gorilla language, staring means you're angry and looking for a fight. Sometimes, gorillas beat their chests when they're cross.

Which monkey has a long nose?

The male proboscis monkey from Borneo has a giant conk, which sometimes droops down below its chin! No one really knows what it's for, but when the monkey makes its loud honking noise through the mangrove trees, its nose straightens out. Maybe the nose makes its call louder, like a loudhailer.

Which primate is always blushing?

The red uakari's face is bright scarlet in sunshine and completely bare of fur. This shy little monkey lives in the Amazon rainforest. It is an expert nutcracker and can even break hard Brazil nut shells.

Amazing!
Primates go bald!

Some chimps and bonobos in central Africa develop bald spots as they get older. The males lose their hair in a perfect triangle shape, and the females lose the lot!

Is it true?
Primates wear glasses.

No. But the spectacled langur of Malaysia looks like it does! Most of its face is covered in dark fur, but the monkey has white circles around its eyes that look like specs.

Which is the punkiest primate?

The cute cotton-top tamarin is the punk rocker of the primate world, with its mohican-style white crest. Lots of primates sport weird hairstyles. The tassel-eared marmoset has tufts of hair above its ears, while the emperor tamarin has a curly white moustache! These tiny monkeys are between 13 and 37 centimetres long, and live mostly in South American tropical forests.

Cotton-top tamarin

Amazing! The tarsier has enormous, goggly eyes. If you had eyes as large in relation to your head, they'd be as big as grapefruits!

Who hangs on for dear life?

The spectral tarsier of Indonesia spends the day sleeping, clinging tightly to a vertical branch. The tips of its fingers are like flattened, sticky pads, and they give it a good grip on its peculiar bed. They also come in handy when the tarsier is hunting for the grubs and insects that it likes to eat.

Is it true?
The aye-aye brings bad luck.

No. But some people living on Madagascar think it does. They believe that if they see one and don't kill it, someone in their village will die. That's one of the reasons why this odd-looking primate is in danger of disappearing.

When do pottos play dead?

When a potto spots a predator (hunter), it sometimes pretends to be dead. It just lets go of its tree and drops to the forest floor. It has another form of defence, too. Its extra-thick neck protects it if it's snatched by a hungry predator.

Which primate has bat-like ears?

The aye-aye's ears are huge and give it amazing hearing. They help it to hear grubs gnawing away under the tree bark. The aye-aye also has a long middle finger, for winkling out its juicy dinner.

What ate 500 figs in a single week?

All lemurs love to come across a tree of juicy figs, but one ruffed lemur once ate about 500 figs in a week. The greedy lemur defended the crop of fruit against any would-be raiders!

Ruffed lemur

Is it true?
Tarsiers have a swivelling head.

Yes. Tarsiers can turn their head half a full circle, like an owl. This is a perfect way to catch an unsuspecting katydid or other flying insect.

Baby orang-utan

Do monkeys and apes eat bananas?

Primates do eat bananas and even peel them first. Fruit, seeds, flowers, shoots, leaves and fungi (types of mushroom) are all perfect primate meals. The orang-utan's favourite snack is the stinky durian fruit, which smells like cheese.

Which primates eat poison?

Lorises eat insects that are so toxic (poisonous) that they would give other animals a heart attack! They sneak up on their prey and grab it with their hands. The golden bamboo lemur even eats young bamboo shoots that contain cyanide, which is a very dangerous poison.

Amazing! Primates chew gum. Many primates, especially marmosets and bushbabies, scrape away the bark of a gum tree to get at the sap. But when it's fresh the gum is liquid, so the animals drink rather than chew.

How do lemur babies get around?

Like all primates, newborn lemur babies cling to their mum's tum as she moves about the forest. As they get bigger and more curious, they have a piggyback, to get a better view.

Is it true?
Only mums look after primate babies.

No. Baby titis are looked after by dad, and young male baboons often borrow a baby. No older male will attack, in case they harm the baby!

Do primate babies drink milk?

Primates are mammals – they give birth to live young and feed them milk. Most primates, including humans, usually have one baby at a time, but marmosets usually have twins.

Baby orang-utan

Amazing! Baby gibbons wear bonnets. When it's born, a baby gibbon has a cap of fur on the top of its head. Just like human babies, the rest of the baby gibbon's body is completely bare!

Why are monkeys so cheeky?

All young monkeys love to play and it's as important as school is for you! This is how they learn the skills they will need when they grow up.

Baby orang-utans

What likes to be by itself?

Most primates live in noisy groups, but there are a few loners. The weasel lemur spends most of its life on its own, although it checks on its neighbours by shouting across the forest. The orang-utan prefers to live on its own too, in the dense jungles of Borneo and Sumatra.

Amazing! Primates have temper tantrums. When a monkey feels cross it sometimes shakes a tree to let off steam, instead of getting angry with the other monkeys, and falling out with its group.

Patas monkeys

Who commands the troop?

The patas monkey is sometimes known as the military monkey. A male leads each group, and makes sure that the 20 females and young are orderly and well-behaved.

Why do primates pick on each other?

Primates often spend much of the day picking through each other's fur. They are looking for tiny biting pests such as ticks, fleas, lice and mites, but grooming is also a way to share smells and bond with each other.

Is it true?
Lemurs drop stink bombs.

Yes. Ring-tailed lemur males have stink fights at the borders of their home range. They load their tail with strong-smelling scent and then swish the smell at their rival!

Chimpanzees

Which is the noisiest primate?

The howler monkey is well-named, because it makes a terrific noise that can be heard up to five kilometres away through the trees. They hold howling competitions with neighbouring groups, to remind each other to keep to their own part of the forest.

Amazing! Monkeys do sentry duty. When a troop is enjoying a feast, one or two animals keep a lookout for predators. They have many different warning calls. For example, they make a certain noise only if a leopard is nearby.

Who grins with fright?

Chimpanzees have a special fear grin. They use it to warn others of danger without making a giveaway noise. Sometimes, when chimps come face to face with a predator, they use the horrible grin to try to frighten it away.

Is it true?
Bushbabies rub wee on their feet!

Yes. And they also rub wee on their hands! It's their way of leaving lots of smelly graffiti on the trees. Every place they've gripped has a scent which says 'we were here'.

Bushbaby

Whose bottom has something to say?

Mandrill

Many primates have brightly-coloured bottoms which are easy to see in the dim forest light. These bottoms tell other members of their own kind where they are. The mandrill's bottom is the most colourful – it's bright blue and red!

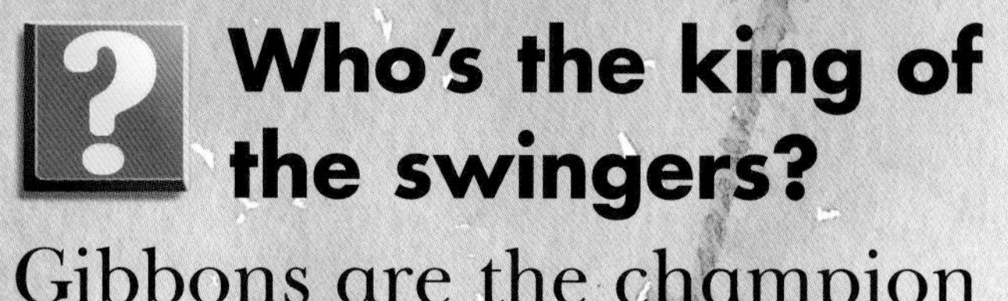

Who's the king of the swingers?

Gibbons are the champion swingers. They have special bones in their wrists and shoulders to give them plenty of swing as they move from tree to tree. These long-armed apes live in the tropical forests of Malaysia and Indonesia.

Amazing! Primates have their own cushions. Many primates, including baboons which spend a lot of time sitting around, have built-in padding on their bottoms.

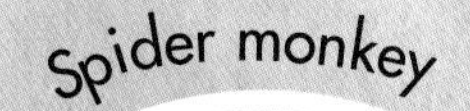

Spider monkey

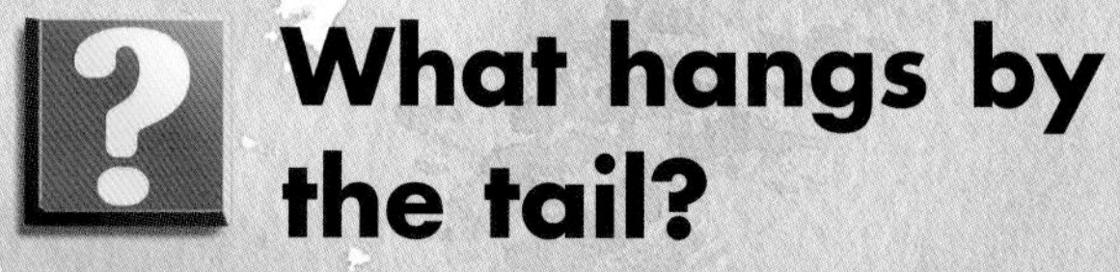

What hangs by the tail?

Woolly monkeys, spider monkeys and howler monkeys all have a bare patch of tail for extra grip. They are the only primates that can support all their weight with the tail and hang upside-down.

Silvery gibbons

Is it true?
Slow lorises really do move slowly.

Yes. Lorises are the most relaxed primates. Unlike their busy monkey cousins, lorises stroll very slowly through the forest in search of food.

Lemur

When are two legs better than four?

Crossing open ground can be a dangerous business with predators about. Lemurs can travel much more quickly on two legs than four. Standing upright also gives them a better view, and frees up their front legs, to pick up food.

Amazing! Chimps take medicine. Chimpanzees sometimes eat plants that don't taste very nice at all, as cures for illness. One herbal remedy is aspilia, which gets rid of tummy upsets and worms.

Which ape uses tools?

Chimps are very clever, and even make simple tools. They sometimes strip a stick of its bark to make a kind of fishing rod that they use to fish for termites. They also use sticks to gather honey so they needn't get too close to the nest and risk a nasty bee sting!

Can apes paint?

Tame chimps and gorillas have been given paints and paper so they can make pictures. Some of the results look like the work of human artists, and foxed a few of the so-called experts who couldn't tell the difference!

Is it true?
Chimps can talk.

No. People have taught chimps to point at symbols and to use sign language, so we know that they are clever enough to understand language. But chimps' vocal cords are unable to produce spoken words like our's.

Who carries a pet stone?

Chimpanzees who live on Mount Tai, in West Africa, use a stone as a nutcracker to smash open the hard shells of the coula nut. There aren't many rocks on the mountain, so each chimp carries around its own favourite stone.

Chimpanzee

Which monkeys climb trees for cash?

In Sumatra, macaques are used to harvest coconuts. They race to the top of the palm and shake off the fruit for money. The monkeys don't keep the cash, but their owners do! Each year, there are competitions to see who are faster, human or monkey coconut-pickers. Of course, the monkeys usually win!

Is it true?
Monkeys and music go together.

Yes. Or at least, they used to. Over 100 years ago street performers called organ-grinders used to travel from town to town with their barrel organs. They used cute capuchin monkeys to collect money from the audience.

Can apes survive in space?

In 1961, a chimp called Ham was sent into space and survived the 16-minute flight. Ham paved the way for people to travel to space.

Which monkeys use computers?

Capuchin monkeys are so clever and easy to train that some have been taught to do all sorts of tasks, including using computers! These helper monkeys live with disabled owners.

Amazing! The pharaohs of ancient Egypt often kept primate pets. The animals had their own servants and were lavished with gifts. Still, their favourites would have been big bowls of fruit, not glittering jewels!

Amazing! Some people think that there are primates yet to be discovered – abominable snowmen! There are many tales of monstrous primates in remote parts of the world, including the yeti from central Asia.

Should people keep primates in zoos?

Primates are happiest in the wild, but zoos do important work. They breed animals that are becoming endangered, such as the golden lion tamarin or the silvery marmoset. Zoos also help people to learn about their ape and monkey cousins. This helps people to understand better why primates should be protected in the wild.

Why are primates in danger?

Not all primates are threatened, but some are. Some, such as the emperor tamarin with its beautiful whiskery moustache, are caught to be sold as pets. Gorillas and orang-utans are in danger because people are destroying their habitat and are also hunting them. There are only about 650 gorillas left in the wild.

Is it true?
People eat chimp and chips.

No. A few apes and monkeys are caught for food, but the biggest threats are the pet trade and the destruction of the places where they live.

Bamboo lemur

Which primate came back from the dead?

Sometimes primates are thought to be extinct, only to re-appear. This happened with the greater bamboo lemur. Most primates are shy and good at hiding. Also, they often live in remote places which are difficult to explore.

Glossary

Ape A tailless primate; a gorilla, orang-utan, chimp, bonobo, gibbon or siamang.

Endangered At risk of dying out.

Extinct No longer alive anywhere on Earth.

Fossil Preserved remains of a creature.

Genes Special instructions inside every cell of a living being, which tell it how to grow.

Grooming Cleaning the fur of ticks and fleas. Primates often do this for each other.

Habitat The type of place where an animal lives in the wild, for example the rainforest.

Helper monkey A monkey which has been bred and trained to help and live with a disabled owner.

Mammal A warm-blooded animal covered in fur which gives birth to live young and feeds its babies on mother's milk.

Monkey Usually a primate with a tail, although not all monkeys have tails.

Predator An animal which hunts other animals for food.

Prey An animal which is hunted by another animal for food.

Primate A group of big-brained mammals, with forward facing eyes, made up of six different groups: lemurs; lorises and galagos; tarsiers; New World monkeys; Old World monkeys; and apes and humans.

Rainforest An evergreen tropical forest where there is heavy rainfall most days.

Index